THIS LITTLE FARMER WENT TO MARKET

By B.G. Hennessy
Illustrated by Mary Ann Fraser

First Edition Project Books - Published by Southwest Human Development
Phoenix, Arizona

For Jane, Lizzie, and Natalie.
- B.G.H.

To all the little farmers who help to put food on our tables.
- M.A.F.

First Edition Project Books
Published by Southwest Human Development
2850 North 24th Street, Phoenix, AZ 85008

USA

First Edition 2021

Library of Congress Control Number: 2021941573
Hennessy, B.G., author
Fraser, Mary Ann, illustrator
Parker-Rock, Michelle, editor
This Little Farmer/ Hennessy and Fraser
32 p.

Summary: "*Little farmers and their families bring the fruits and vegetables that they grew on their farms to the farmers market in the city. At the market, the little farmers help to set up stands and prepare for the hungry shoppers.*"
- Provided by publisher.

Manufactered in China by PRC Printing
ISBN 978-0-578-93291-0
10 9 8 7 6 5 4 3 2 1

Design by Michael Hale
Text set in Californian FB with initial caps in Berhard MT

This little farmer goes to market.

That little farmer goes too.

This farmer family has one stand.

That farmer family has two.

And this little farmer helps on their farm all day long.

This farmer's stand sells peaches.

That farmer's stand sells beans.

This farmer's stand sells snap peas.

That farmer's stand sells greens.

And this little farmer helps on his farm all day long.

This farmer's stand sells berries.

That farmer's stand sells treats.

This farmer's stand sells melons.

That farmer's stand sells beets.

And this little farmer helps on her farm all day long.

This farmer's stand sells parsnips.
That farmer's stand sells jams.
This farmer's stand sells tomatoes.
That farmer's stand sells yams.

And this little farmer helps on his farm all day long.

Plant the seeds row by row.

Day by day they sprout and grow.

Tilling, hoeing, raking too.

Watering, weeding, lots to do.

Parsley, rhubarb, pumpkin vines.

Carrots, turnips, harvest time.

Apples, peaches, pears and plums.

Grapes and cherries, yum, yum, yum.

Peppers, onions, yellow squash.
Zucchini, mushrooms, McIntosh.

This little shopper went to market.

That little shopper went too.

This little shopper filled one bag.

That little shopper filled two.

FARMERS MARKET
Sunny Hills
FARM

And these little shoppers
said, “mmm, mmm, mmm,”
all the way home.

Farm Facts

There are many different types of farms.

Farms that grow crops like vegetables, fruits, and grains are called *arable* farms. *Arable* means that the soil can be plowed and planted. Many farms grow more than one crop at a time.

Most fruits and nuts grow on trees on farms called *orchards*.

Grapes grow on a farm called a *vineyard*.

Have you ever been to a *pumpkin patch* or a *tree nursery*? Those are farms too.

Some farms raise animals.

A *dairy farm* raises cows that can be milked.

A *cattle ranch* raises animals for meat.

A *poultry farm* raises chickens, ducks, or turkeys for eggs and meat.

Visit a farm near you.
Some farms even let you pick your own fruits and vegetables.

Hidden Pictures

There are animals and insects hidden in the pictures. Can you find them all?

 Goat

 Snail

 Butterfly

 Dog

 Rabbit

 Bee

 Ladybug

 Crow

 Cat

 Chicken

 Snake

 Piglet

 Worm

 Owl

 Chick

A Farmer's Toolshed

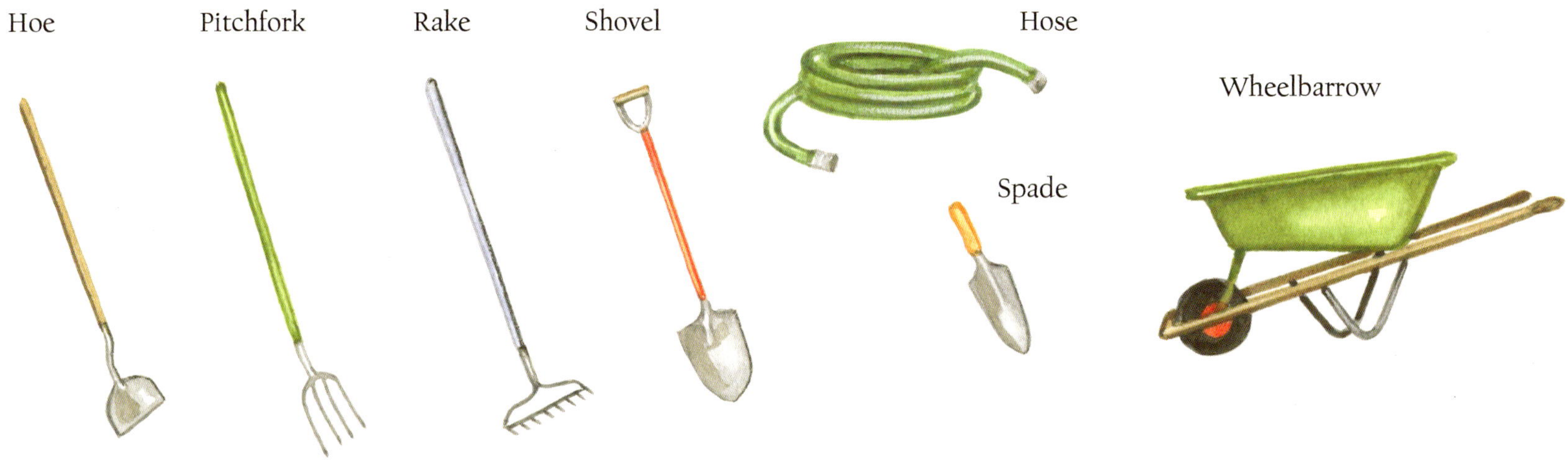

What does a farm that grows crops need?

A hard working farmer.

Seeds. Most plants grow from seeds.

A good source of water. Not too much and not too little.

The soil cannot be too rocky or sandy. Good soil has lots of nutrients that feed the plants. Nutrients are like vitamins.

The right climate. It cannot be too cold or too hot. Different plants grow in different zones. Every plant needs sunlight.

If you would like to learn more, go to...

Remember at your next meal, a lot of the food we eat comes from a farm.